I0492790

The information provided in this book is designed to provide helpful information on the Pick 4. This book is for entertainment purposes only. No part or parts of this book shall be copied or used without the sole consent of the author. Guides or work outs are provided for informational purposes only and do not constitute endorsement of any lottery websites or other sources. Readers should be aware that the author is not employed by any gaming companies now or in the past.

Winning the pick 4

Lottery Guide

The numbers in this guide are produced in the form of a circle. When you find a match in this guide then you need to play only that circle until it hits. A circle is a line of numbers . I am famous for my circles because they are more cost effective than chasing numbers . The numbers will hit its just a matter of patience. I have not put the numbers in order because numbers do not fall in order. You may have to look for the pick 4 number that fell through a few pages but if your like so many that play pick 4 numbers and never hit then this is the absolute best book for you. This book is good for anywhere the Pick 4 is played .

Example if 0135 falls in the pick 4 only play the circle you find associated with that until another number from the circle hits.

0135 0139 0159 0199 0359 0399 0599 1359 1399 1599 3599 in this example you can play them for 50 cents each draw for a total of $5.50 .My circle numbers are meant to be low cost playing solutions.

I hope you find the pick 4 circles an asset to your everyday plays . Please also see my pick 3 circle book as well for your pick 3 daily plays.

Here are the pick 4 circle numbers .

0037 0038 0078 0088 0378 0388 0788 3788

0234 0237 0238 0247 0248 0278 0347 0348 0378 0478
2347 2348 2378 2478 3478

0134 0139 0149 0199 0349 0399 0499 1349 1399 1499
3499

0136 0139 0169 0199 0369 0399 0699 1369 1399 1699
3699

2345 2346 2347 2356 2357 2367 2456 2457 2467 2567
3456 3457 3467 3567 4567

0012 0013 0016 0023 0026 0036 0123 0126 0136 0236
1236

0011 0018 0111 0118 1118

0112 0114 0122 0124 0224 1122 1124 1224

0156 0166 0566 0666 1566 1666 5666

0233 0234 0333 0334 2333 2334 3334

2467 2477 2677 2777 4677 4777 6777

1113 1119 1133 1139 1339

0006 0007 0008 0067 0068 0078 0678

0012 0013 0019 0023 0029 0039 0123 0129 0139 0239
1239

0134 0136 0137 0146 0147 0167 0346 0347 0367 0467
1346 1347 1367 1467 3467

0135 0136 0137 0156 0157 0167 0356 0357 0367 0567
1356 1357 1367 1567 3567

1677 1777 6777 7777

1556 1557 1558 1567 1568 1578 1678 5567 5568 5578
5678

0115 0117 0119 0157 0159 0179 0579 1157 1159 1179
1579

0066 0067 0069 0079 0667 0669 0679 6679

0034 0035 0039 0045 0049 0059 0345 0349 0359 0459
3459

0237 0238 0277 0278 0377 0378 0778 2377 2378 2778
3778

2677 2678 2688 2778 2788 6778 6788 7788

1223 1227 1237 1277 1377 2237 2277 2377

1134 1136 1144 1146 1344 1346 1446 3446

0003 0008 0009 0038 0039 0089 0389

2344 2345 2346 2356 2445 2446 2456 3445 3446 3456
4456

0133 0137 0177 0337 0377 1337 1377 3377

0124 0126 0127 0146 0147 0167 0246 0247 0267 0467
1246 1247 1267 1467 2467

1445 1449 1459 1499 1599 4459 4499 4599

0246 0247 0249 0267 0269 0279 0467 0469 0479 0679
2467 2469 2479 2679 4679

2338 2339 2388 2389 2889 3388 3389 3889

2235 2237 2238 2257 2258 2278 2357 2358 2378 2578
3578

0678 0679 0688 0689 0788 0789 0889 6788 6789 6889
7889

0123 0128 0129 0138 0139 0189 0238 0239 0289 0389
1238 1239 1289 1389 2389

1256 1257 1267 1277 1567 1577 1677 2567 2577 2677
5677

0034 0037 0038 0047 0048 0078 0347 0348 0378 0478
3478

0333 0336 3333 3336

4477 4478 4479 4489 4778 4779 4789 7789

0245 0249 0255 0259 0455 0459 0559 2455 2459 2559
4559

1345 1349 1359 1399 1459 1499 1599 3459 3499 3599
4599

0146 0148 0166 0168 0466 0468 0668 1466 1468 1668
4668

0077 0078 0079 0089 0778 0779 0789 7789

0367 0369 0377 0379 0677 0679 0779 3677 3679 3779
6779

0144 0145 0147 0157 0445 0447 0457 1445 1447 1457
4457

0011 0016 0017 0067 0116 0117 0167 1167

1249 1299 1499 1999 2499 2999 4999

0067 0069 0079 0099 0679 0699 0799 6799

2367 2368 2378 2388 2678 2688 2788 3678 3688
3788 6788

1345 1347 1357 1377 1457 1477 1577 3457 3477
3577 4577

3555 3557 3558 3578 5557 5558 5578

3345 3346 3356 3366 3456 3466 3566 4566

0017 0019 0079 0099 0179 0199 0799 1799

0188 0189 0199 0889 0899 1889 1899 8899

1225 1226 1256 1266 1566 2256 2266 2566

0556 0558 0559 0568 0569 0589 0689 5568 5569
5589 5689

2256 2257 2258 2267 2268 2278 2567 2568 2578
2678 5678

0333 0334 0336 0346 3334 3336 3346

1266 1267 1269 1279 1667 1669 1679 2667 2669
2679 6679

0245 0247 0249 0257 0259 0279 0457 0459 0479
0579 2457 2459 2479 2579 4579

0112 0114 0116 0124 0126 0146 0246 1124 1126
1146 1246

1346 1348 1368 1388 1468 1488 1688 3468 3488
3688 4688

1247 1248 1249 1278 1279 1289 1478 1479 1489
1789 2478 2479 2489 2789 4789

0077 0078 0079 0089 0778 0779 0789 7789

2467 2469 2477 2479 2677 2679 2779 4677 4679
4779 6779

2338 2339 2388 2389 2889 3388 3389 3889

0356 0359 0366 0369 0566 0569 0669 3566 3569
3669 5669

1249 1299 1499 1999 2499 2999 4999

0014 0018 0019 0048 0049 0089 0148 0149 0189
0489 1489

0138 0139 0189 0199 0389 0399 0899 1389 1399
1899 3899

1456 1458 1459 1468 1469 1489 1568 1569 1589
1689 4568 4569 4589 4689 5689

0012 0014 0016 0024 0026 0046 0124 0126 0146
0246 1246

1235 1237 1239 1257 1259 1279 1357 1359 1379
1579 2357 2359 2379 2579 3579

0067 0069 0079 0099 0679 0699 0799 6799

1244 1246 1248 1268 1446 1448 1468 2446 2448
2468 4468

1234 1235 1239 1245 1249 1259 1345 1349 1359
1459 2345 2349 2359 2459 3459

1344 1345 1347 1357 1445 1447 1457 3445 3447
3457 4457

0114 0119 0149 0199 0499 1149 1199 1499

2347 2348 2349 2378 2379 2389 2478 2479 2489
2789 3478 3479 3489 3789 4789

4456 4458 4459 4468 4469 4489 4568 4569 4589
4689 5689

0024 0026 0028 0046 0048 0068 0246 0248 0268
0468 2468

0455 0456 0457 0467 0556 0557 0567 4556 4557
4567 5567

0233 0237 0239 0279 0337 0339 0379 2337 2339
2379 3379

2344 2444 3444 4444

0124 0126 0128 0146 0148 0168 0246 0248 0268
0468 1246 1248 1268 1468 2468

0355 0357 0359 0379 0557 0559 0579 3557 3559
3579 5579

1223 1225 1226 1235 1236 1256 1356 2235 2236
2256 2356

0234 0236 0238 0246 0248 0268 0346 0348 0368
0468 2346 2348 2368 2468 3468

0247 0248 0249 0278 0279 0289 0478 0479 0489
0789 2478 2479 2489 2789 4789

5668 5669 5688 5689 5889 6688 6689 6889

0122 0123 0128 0138 0223 0228 0238 1223 1228
1238 2238

0022 0028 0029 0089 0228 0229 0289 2289

2334 2335 2339 2345 2349 2359 2459 3345 3349
3359 3459

5666 5669 5699 6669 6699

1456 1458 1459 1468 1469 1489 1568 1569 1589
1689 4568 4569 4589 4689 5689

1344 1345 1347 1357 1445 1447 1457 3445 3447
3457 4457

0245 0247 0248 0257 0258 0278 0457 0458 0478
0578 2457 2458 2478 2578 4578

2337 2339 2379 2399 2799 3379 3399 3799

0155 0158 0555 0558 1555 1558 5558

0257 0258 0259 0278 0279 0289 0578 0579 0589
0789 2578 2579 2589 2789 5789

0666 0668 0688 6668 6688

0046 0048 0068 0088 0468 0488 0688 4688

0344 0345 0349 0359 0445 0449 0459 3445 3449
3459 4459

0335 0339 0355 0359 0559 3355 3359 3559

4488 4489 4888 4889 8889

1457 1458 1459 1478 1479 1489 1578 1579 1589
1789 4578 4579 4589 4789 5789

1245 1247 1248 1257 1258 1278 1457 1458 1478
1578 2457 2458 2478 2578 4578

3378 3379 3389 3399 3789 3799 3899 7899

0012 0014 0018 0024 0028 0048 0124 0128 0148
0248 1248

1245 1247 1248 1257 1258 1278 1457 1458 1478
1578 2457 2458 2478 2578 4578

4456 4457 4459 4467 4469 4479 4567 4569 4579
4679 5679

1245 1248 1249 1258 1259 1289 1458 1459 1489
1589 2458 2459 2489 2589 4589

0224 0226 0228 0246 0248 0268 0468 2246 2248
2268 2468

1266 1267 1268 1278 1667 1668 1678 2667 2668
2678 6678

1456 1457 1459 1467 1469 1479 1567 1569 1579
1679 4567 4569 4579 4679 5679

1237 1238 1278 1288 1378 1388 1788 2378 2388
2788 3788

1134 1136 1144 1146 1344 1346 1446 3446

0127 0128 0177 0178 0277 0278 0778 1277 1278
1778 2778

0166 0167 0169 0179 0667 0669 0679 1667 1669
1679 6679

3488 3489 3888 3889 4888 4889 8889

2357 2358 2359 2378 2379 2389 2578 2579 2589
2789 3578 3579 3589 3789 5789

2256 2257 2259 2267 2269 2279 2567 2569 2579 2679 5679

0045 0048 0055 0058 0455 0458 0558 4558

0345 0346 0348 0356 0358 0368 0456 0458 0468 0568 3456
3458 3468 3568 4568

1122 1124 1128 1148 1224 1228 1248 2248

0133 0135 0139 0159 0335 0339 0359 1335 1339 1359 3359
0013 0017 0033 0037 0133 0137 0337 1337

4456 4457 4466 4467 4566 4567 4667 5667

1256 1259 1269 1299 1569 1599 1699 2569 2599
2699 5699

0268 0269 0288 0289 0688 0689 0889 2688 2689
2889 6889

0145 0148 0158 0188 0458 0488 0588 1458 1488
1588 4588

2388 2389 2888 2889 3888 3889 8889

0134 0137 0138 0147 0148 0178 0347 0348 0378
0478 1347 1348 1378 1478 3478

1225 1227 1255 1257 1557 2255 2257 2557

3789 3799 3899 3999 7899 7999 8999

1246 1248 1249 1268 1269 1289 1468 1469 1489
1689 2468 2469 2489 2689 4689

1346 1348 1366 1368 1466 1468 1668 3466 3468
3668 4668

0234 0236 0244 0246 0344 0346 0446 2344 2346
2446 3446

0146 0149 0169 0199 0469 0499 0699 1469 1499
1699 4699

0125 0126 0129 0156 0159 0169 0256 0259 0269
0569 1256 1259 1269 1569 2569

0011 0016 0018 0068 0116 0118 0168 1168

0556 0557 0559 0567 0569 0579 0679 5567 5569
5579 5679

1112 1113 1117 1123 1127 1137 1237

0235 0237 0238 0257 0258 0278 0357 0358 0378
0578 2357 2358 2378 2578 3578

2678 2679 2689 2699 2789 2799 2899 6789 6799
6899 7899

0247 0248 0278 0288 0478 0488 0788 2478 2488
2788 4788

3667 3668 3678 3688 3788 6678 6688 6788

0112 0116 0119 0126 0129 0169 0269 1126 1129
1169 1269

0039 0099 0399 0999 3999

0033 0038 0039 0089 0338 0339 0389 3389

0056 0057 0058 0067 0068 0078 0567 0568 0578
0678 5678

0333 0334 0335 0345 3334 3335 3345

0117 0119 0177 0179 0779 1177 1179 1779

1344 1346 1349 1369 1446 1449 1469 3446 3449
3469 4469

0036 0037 0038 0067 0068 0078 0367 0368 0378 0678 3678

2477 2478 2479 2489 2778 2779 2789 4778 4779 4789 7789

4557 4558 4559 4578 4579 4589 4789 5578 5579 5589 5789

0056 0058 0059 0068 0069 0089 0568 0569 0589 0689 5689

0113 0115 0118 0135 0138 0158 0358 1135 1138 1158 1358

1267 1269 1279 1299 1679 1699 1799 2679 2699 2799 6799

0112 0116 0118 0126 0128 0168 0268 1126 1128 1168 1268

0667 0668 0669 0678 0679 0689 0789 6678 6679 6689 6789

1136 1137 1166 1167 1366 1367 1667 3667

1233 1235 1255 1335 1355 2335 2355 3355

0057 0058 0059 0078 0079 0089 0578 0579 0589 0789 5789

0034 0035 0037 0045 0047 0057 0345 0347 0357 0457 3457

1244 1245 1249 1259 1445 1449 1459 2445 2449 2459 4459

0348 0349 0389 0399 0489 0499 0899 3489 3499 3899 4899

0245 0246 0255 0256 0455 0456 0556 2455 2456 2556 4556

0447 0448 0449 0478 0479 0489 0789 4478 4479 4489 4789

2337 2377 2777 3377 3777

2344 2346 2347 2367 2446 2447 2467 3446 3447 3467 4467

0366 0367 0369 0379 0667 0669 0679 3667 3669 3679 6679

1456 1457 1458 1467 1468 1478 1567 1568 1578 1678 4567 4568 4578 4678 5678

0255 0256 0258 0268 0556 0558 0568 2556 2558 2568 5568

2456 2457 2459 2467 2469 2479 2567 2569 2579 2679 4567 4569 4579 4679 5679

1244 1245 1248 1258 1445 1448 1458 2445 2448 2458 4458

0346 0348 0366 0368 0466 0468 0668 3466 3468 3668 4668

3477 3478 3488 3778 3788 4778 4788 7788

3467 3468 3469 3478 3479 3489 3678 3679 3689 3789 4678 4679 4689 4789 6789

1344 1345 1349 1359 1445 1449 1459 3445 3449 3459 4459

3556 3558 3559 3568 3569 3589 3689 5568 5569 5589 5689

0134 0138 0148 0188 0348 0388 0488 1348 1388 1488 3488

0257 0259 0277 0279 0577 0579 0779 2577 2579 2779 5779

0248 0249 0288 0289 0488 0489 0889 2488 2489 2889 4889

2336 2339 2366 2369 2669 3366 3369 3669

0046 0049 0069 0099 0469 0499 0699 4699

0157 0158 0178 0188 0578 0588 0788 1578 1588 1788 5788

1233 1236 1237 1267 1336 1337 1367 2336 2337 2367 3367

2335 2337 2338 2357 2358 2378 2578 3357 3358 3378 3578

2345 2346 2348 2356 2358 2368 2456 2458 2468 2568 3456 3458 3468 3568 4568

0047 0049 0077 0079 0477 0479 0779 4779

0126 0127 0167 0177 0267 0277 0677 1267 1277 1677 2677

0337 0338 0378 0388 0788 3378 3388 3788

0245 0246 0249 0256 0259 0269 0456 0459 0469 0569 2456 2459 2469 2569 4569

1255 1256 1555 1556 2555 2556 5556

1677 1679 1777 1779 6777 6779 7779

0223 0224 0227 0234 0237 0247 0347 2234 2237 2247 2347

1348 1349 1389 1399 1489 1499 1899 3489 3499
3899 4899

2345 2347 2348 2357 2358 2378 2457 2458 2478
2578 3457 3458 3478 3578 4578

0122 0126 0166 0226 0266 1226 1266 2266

2445 2447 2448 2457 2458 2478 2578 4457 4458
4478 4578

0234 0238 0239 0248 0249 0289 0348 0349 0389
0489 2348 2349 2389 2489 3489

0237 0238 0278 0288 0378 0388 0788 2378 2388
2788 3788

0136 0137 0138 0167 0168 0178 0367 0368 0378
0678 1367 1368 1378 1678 3678

1245 1249 1259 1299 1459 1499 1599 2459 2499
2599 4599

0344 0347 0377 0447 0477 3447 3477 4477

0467 0468 0477 0478 0677 0678 0778 4677 4678
4778 6778

0466 0467 0469 0479 0667 0669 0679 4667 4669
4679 6679

0123 0124 0125 0134 0135 0145 0234 0235 0245
0345 1234 1235 1245 1345 2345

0266 0267 0268 0278 0667 0668 0678 2667 2668
2678 6678

0013 0015 0018 0035 0038 0058 0135 0138 0158
0358 1358

4688 4689 4888 4889 6888 6889 8889

0246 0247 0266 0267 0466 0467 0667 2466 2467 2667 4667

1356 1357 1359 1367 1369 1379 1567 1569 1579 1679 3567 3569 3579 3679 5679

3366 3369 3399 3669 3699 6699

1233 1237 1239 1279 1337 1339 1379 2337 2339 2379 3379

3466 3468 3666 3668 4666 4668 6668

0035 0038 0039 0058 0059 0089 0358 0359 0389 0589 3589

4445 4446 4448 4456 4458 4468 4568

1123 1127 1137 1177 1237 1277 1377 2377

0046 0049 0066 0069 0466 0469 0669 4669

0015 0016 0019 0056 0059 0069 0156 0159 0169 0569 1569

1239 1299 1399 1999 2399 2999 3999

0023 0024 0026 0034 0036 0046 0234 0236 0246 0346 2346

0244 0246 0248 0268 0446 0448 0468 2446 2448 2468 4468

0234 0239 0244 0249 0344 0349 0449 2344 2349 2449 3449

0335 0339 0355 0359 0559 3355 3359 3559

3345 3348 3355 3358 3455 3458 3558 4558

0222 0226 2222 2226

3345 3347 3348 3357 3358 3378 3457 3458 3478 3578 4578

2336 2337 2339 2367 2369 2379 2679 3367 3369 3379 3679

0112 0113 0114 0123 0124 0134 0144 0234 0244 0344 0444 1123 1124 1134 1144 1234 1244 1344 1444 2344 2444 3444 4444

1336 1339 1366 1369 1669 3366 3369 3669

0334 0336 0339 0346 0349 0369 0469 3346 3349 3369 3469

1226 1227 1229 1267 1269 1279 1679 2267 2269 2279 2679

0034 0039 0049 0099 0349 0399 0499 3499

2246 2248 2249 2268 2269 2289 2468 2469 2489 2689 4689

0445 0446 0449 0456 0459 0469 0569 4456 4459 4469 4569

1123 1124 1129 1134 1139 1149 1234 1239 1249 1349 2349

0122 0126 0127 0167 0226 0227 0267 1226 1227 1267 2267

0446 0447 0449 0467 0469 0479 0679 4467 4469 4479 4679

0044 0045 0047 0057 0445 0447 0457 4457

0377 0378 0379 0389 0778 0779 0789 3778 3779 3789 7789

0235 0236 0239 0256 0259 0269 0356 0359 0369
0569 2356 2359 2369 2569 3569

0138 0139 0189 0199 0389 0399 0899 1389 1399
1899 3899

0045 0046 0047 0056 0057 0067 0456 0457 0467
0567 4567

0446 0448 0466 0468 0668 4466 4468 4668

1117 1118 1119 1178 1179 1189 1789

0156 0157 0167 0177 0567 0577 0677 1567 1577
1677 5677

0023 0028 0038 0088 0238 0288 0388 2388

1133 1134 1138 1148 1334 1338 1348 3348

0445 0447 0455 0457 0557 4455 4457 4557

0234 0237 0239 0247 0249 0279 0347 0349 0379
0479 2347 2349 2379 2479 3479

3346 3349 3369 3399 3469 3499 3699 4699

2233 2237 2239 2279 2337 2339 2379 3379

0127 0128 0178 0188 0278 0288 0788 1278 1288
1788 2788

0122 0124 0125 0145 0224 0225 0245 1224 1225
1245 2245

2468 2469 2488 2489 2688 2689 2889 4688 4689
4889 6889

0123 0124 0125 0134 0135 0145 0234 0235 0245
0345 1234 1235 1245 1345 2345

0014 0015 0017 0045 0047 0057 0145 0147 0157
0457 1457

1667 1669 1679 1699 1799 6679 6699 6799

0046 0048 0066 0068 0466 0468 0668 4668

2567 2569 2577 2579 2677 2679 2779 5677 5679
5779 6779

2455 2456 2459 2469 2556 2559 2569 4556 4559
4569 5569

1234 1238 1244 1248 1344 1348 1448 2344 2348
2448 3448

1355 1357 1359 1379 1557 1559 1579 3557 3559
3579 5579

1114 1116 1144 1146 1446

1222 1226 1227 1267 2226 2227 2267

1123 1129 1139 1199 1239 1299 1399 2399

0127 0128 0129 0178 0179 0189 0278 0279 0289
0789 1278 1279 1289 1789 2789

1244 1248 1249 1289 1448 1449 1489 2448 2449
2489 4489

1357 1358 1377 1378 1577 1578 1778 3577 3578
3778 5778

www.ingramcontent.com/pod-product-compliance
Lightning Source LLC
Chambersburg PA
CBHW071205220526
45468CB00003B/1168